Japheth Can Count

Counting And Colors

Japheth Can Count
Counting And Colors

Author Rai White

Japheth Can Count

Counting And Colors

A Children's Book

By Author Rai White

More Books By Rai White

- ❖ Nigel's Choice
- ❖ Before And After I Do
- ❖ Real Love Is….
- ❖ Real Love Is….2: The 25th Year
- ❖ From Rags To Stitches: My Memoir

Japheth Can Count

Counting And Colors

Published by Raynell White

Copyright © 2016 Raynell White

Cover Artist/Illustrator William W. White Jr.

Printed In United States of America

ISBN-10 1535473983

ISBN-13 978-1535473989

Contact information about this book or any other books by

Rai White: ithinkican1963@yahoo.com

Purchasing Info: amzn.to/2ok5aJj

Instagram: @hermosa_perspectiva

vimeo.com/235431767

Facebook: @raiwhiteauthor

Acknowledgements

This book is dedicated to my one and only three-year-old grandchild, Japheth Omar. He's been counting and identifying some colors since he was between 1 ½ and 2 years old. My family and I love our little prince so much, and I've desired to write a book about him since he was an infant. However, my busy schedule wouldn't allow for it, at the time. Nevertheless, this is a start. One day soon, Japheth will look at these pictures of himself and chuckle at how little he once was and will be amazed at how much he's grown. They grow up fast.

Japheth Omar at 1 ½ year

Japheth Omar- 3 years old

Preface

I've heard professionals say that repetition is a part of the learning process for young children. Therefore, I've prepared the text in this book so that young children will repeat the numbers and colors. They will also learn to identify different objects. I hope this will result in them having a fun and easy learning experience with this book.

One

One elephant

What color is the elephant?

One gray elephant

Japheth saw a gray elephant at the zoo.

2

One-two

Two birds

What color are the birds?

The birds are black.

Two black birds

Japheth saw two black birds standing next to each other.

One-two-three

Three bananas

What color are the bananas?

The bananas are yellow.

Three yellow bananas

Japheth eats a yellow banana for his snack.

4

One-two-three-four

Four pine cones

What color are the pine cones?

The pine cones are brown.

Four brown pine cones

Japheth saw lots of brown pine cones on the ground.

5

One-two-three-four-five

Five pigs

What color are the pigs?

The pigs are pink.

Japheth saw some pink pigs on the farm.

6

One-two-three-four-five-six

Six airplanes

What color are the airplanes?

The airplanes are blue.

Six blue airplanes

Japheth flew on an airplane high in the sky, for his birthday.

When Japheth got off the airplane, the nice pilot gave him a pair of golden, wings.

One-two-three-four-five-six-seven

Seven cars

What color are the cars?

The cars are orange.

Seven orange cars

Japheth plays with his cars every day.

8

One-two-three-four-five-six-seven-eight

Eight trees

What color are the trees?

The trees are green.

Japheth likes looking at the trees beside the road, when he's riding in his car.

9

One-two-three-four-five-six-seven-eight-nine

Nine flowers

What color are the flowers?

The flowers are purple.

Nine purple flowers

Japheth picks fresh flowers for his grandma, when he visits her and grandpa. His grandma says that he is such a gentleman.

10

One-two-three-four-five-six-seven-eight-nine-ten

Ten wagons

What color are the wagons?

The wagons are red.

Ten red wagons

Japheth has a red wagon to play and ride in.

His mom pulls the wagon when she takes Japheth for a ride.

COLORS

Gray pencil

Black book bag

Yellow duck

Brown bear

Pink pencil

Blue bike

Orange pumpkin

Green tree

Purple flowers

Red car

COUNTING

One-two-three-four-five-six-seven-eight-nine-ten

10 apples

A B C

A-apple

B-ball

C-cat

A

Apple

B

Ball

C

Cat

I trust that this book will be the beginning of a fun learning experience, for many children who are 1 ½ and up. As a mother of four bright young adults, and as a grandmother, I know that reading is fundamental.

~Author Rai White

www.ingramcontent.com/pod-product-compliance
Lightning Source LLC
Chambersburg PA
CBHW050400180526
45159CB00005B/2098